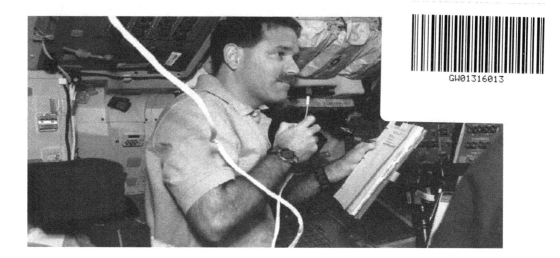

Why We Explore

Welcome to "Risk and Exploration—Earth, Sea, and the Stars." Today's session is entitled "Why We Explore," but I'm hoping that, mostly, we can make it a dialogue, up close and personal. I'm John Grunsfeld. I'm the NASA chief scientist and an astronaut.

I think we have started getting into the discussions on risk and exploration, into some of the thorny questions about how do we make decisions. How do we use our judgment? How do we, as institutional managers of a public institution, make decisions on behalf of the American people, and with oversight of the Congress, that can stand the test of time, without being so risk averse that we don't do anything interesting?

There's a couple of things I'd like to show this morning that are personal, that are professional as chief scientist, and then, representing the Agency, and then, looking forward.

I think we'd be remiss in all of this discussion if we avoided the topic of why we're not sending a Space Shuttle back to the Hubble to service it. So I'll address that in a second.

One of our favorite cartoons shows a Conestoga wagon heading across the Great Plains. And the title reads "Alarmed by the many dangers, the pioneers abandoned westward

John Grunsfeld
NASA Chief Scientist and Astronaut

Astronaut, astrophysicist, and mountaineer, John M. Grunsfeld served as NASA's chief scientist from 2003–2004. Grunsfeld is a veteran of four Space Shuttle flights. In 1999 and 2002 he took part in a total of five successful spacewalks to upgrade the Hubble Space Telescope.

A native of Chicago, Grunsfeld received a bachelor's degree in physics from the Massachusetts Institute of Technology in 1980. He earned a masters degree and a doctorate in physics from the University of Chicago in 1984 and 1988, respectively. Grunsfeld was selected as a NASA astronaut in 1992. His first flight assignment came in 1995 on board the Space Shuttle *Endeavour* on STS-67. In 1997, Grunsfeld served as flight engineer for the Space Shuttle *Atlantis* during STS-81 and a 10-day mission to Russia's *Mir* space station. He has logged over 45 days in space, including 37 hours and 32 minutes working outside the Space Shuttle.

RISK AND EXPLORATION: EARTH, SEA AND THE STARS

OPENING PHOTO:
Astronaut John M. Grunsfeld, mission specialist, looks over a flight plan on Space Shuttle Discovery's flight deck while communicating with ground controllers.
(NASA Image # S103-E-5016)

exploration except for a series of unmanned prairie-probe vehicles." You know, I think many people have summed up succinctly why humans explore. Because we want to go. In the face of danger, but managed risk. I am absolutely positive that our outward expansion from the cradle that planet Earth is, will not be one of strictly unmanned probes, but we will be heading out across the prairies. Why [do] we explore? Right now, Spirit at Columbia Hills [Mars] is poised to look over those hills and see what's beyond.

I'm an explorer who is trained by a group called the National Outdoor Leadership School, and we're privileged to have John Gans, the director of the National Outdoor Leadership School, here. I went there in high school and it was to learn to be a better risk manager, a better leader in the outdoors and, hopefully, not to be reckless like most teenagers.

But my interest in exploration was largely driven through the pages of *National Geographic*, through the movies of Jacques Cousteau, while growing up on the south side of Chicago, that I was able to explore vicariously. But I wanted to go. I have a passion for exploration, and I have a weakness. When I see something like Columbia Hills, I have a need to look over that hill. And it's a real challenge.

It's a real challenge because you set limits for yourself. And as mountaineers, we set limits for ourselves. We have to summit by a certain time so we can make it back safely. And I'm constantly torn, wanting to go further, especially when I'm on professional travel and I take a day off to go hike. I say, "Well, I only have one day and I'll go this far." And I get that far and I look forward and I say, "Boy, I've got to go a little further."

So, that's what we're doing with Spirit and Opportunity on Mars now. We have the opportunity to go further because the rovers are still running, they're still doing great. You know, we had a 90- or a 120-day mission and we're well beyond that now, and we have hope they can go much further.

This May and June, I had the opportunity to try and climb a little hill in Alaska called Denali—Mt. McKinley. It's 6,157 meters, 20,320 feet tall. This is a serious expedition. It's not quite the kind of thing that Ed Viesturs does, but it's, I think, comparable in many ways.

It's at 63 degrees north latitude. That makes it perhaps the coldest mountain on planet Earth. You start out already basically in the Arctic. Its conditions on the summit are comparable to Everest in winter. The Alaska Range is a large landmass that extends up out of plains, basically, a few hundred feet in altitude. It sees the full brunt of arctic weather. And, so, it seemed like an appropriate challenge.

Now, in order to do this as an astronaut—and I see Colonel Cabana in the audience—this was my third try. The first time I tried, as an astronaut, I felt compelled to write a mission statement and a risk-mitigation statement that I submitted to my boss, Colonel Cabana, then chief of the astronaut office, so that I could get permission to go, so to speak. Even though it was personal leave.

That's the way I view risk management on this climb: you have my crew notebook with checklists. And I think I'm the only mountaineer I know who goes up with checklists and says, "Okay." And part of that was, I recognized that at high altitude I will be hypoxic, I will make mistakes. And this was one of my mechanisms to prevent myself from making mistakes. I still made lots of mistakes. I think back now and I think, "It's in the checklist; how could I have missed it?"

But it's one mechanism of risk mitigation that we use very often in the space business, because the line between life and death is so fine. We heard in this talk something that I think is very characteristic: the farther you go from base camp, if the smiles get bigger, you have the right team. And we lived, basically, on a glacier for 23 days. All of our water came from melting snow.

Just a great experience. No cell phones, no beepers, no Blackberrys, really just existing in a very primal way, but with the aid of high technology, and that's something I think is part of the real spirit of exploration, that trying to go to the next hill. And I got up to the top and I looked back and I waved at Dave Schuman, who is another NASA Headquarters employee. I said, "Dave," and I had to yell.

> I HAVE A PASSION FOR EXPLORATION, AND I HAVE A WEAKNESS. WHEN I SEE SOMETHING LIKE COLUMBIA HILLS, I HAVE A NEED TO LOOK OVER THAT HILL. AND IT'S A REAL CHALLENGE.

I said, "Dave, I have bad news." And he was thinking "Oh, no." We thought we were close to the summit—we'd been on the summit ridge for about two hours climbing up from something called the "Football Field." And it's tedious. What Ed Viesturs said is right. You take a step, you breathe a bunch of times, and you take a step. And every time you stop to breathe, you look forward to see how much longer it is. And, very often, you don't see the top, you know, [you have to] climb another ridge. And I said, "I have bad news. There's no place else to go but down."

I was actually worried about sort of an anticlimactic feeling. This was my third try, and I just couldn't believe I was actually standing on the highest point in North America—just an unbelievable feeling. I was half laughing— my climbing buddies say hysterically—and half crying. I just couldn't believe it. So we had three NASA employees on the summit of North America on June 7th of this year, 91 years after the first ascent.

A lot of people have climbed to the top of this mountain—about 12,000. About one out of a hundred perish in this. My risk management plan was to go through a book called *Accidents in North American Mountaineering*. It's published every year. Just the fact that a book like this is published means that mountaineers are very sensitive to this issue of risk and that we try and learn from others' mistakes.

I went through basically every mountaineering accident on Mt. McKinley and in the Alaska Range from 1969 to about 1992 and came up with common causes, behaviors that led to those accidents, and then asked myself, how can I avoid those behaviors? And, so, that was also in my little notebook. And I'd review that every night and then review it with the team. "Okay, we're not going to do this. We're always going to stay roped, no matter where we are. We're always going to carry an ice axe."

We pretty much beat it away so that if you do the statistics, it became more like 1 in 10,000. And one of the things that people think about mountaineering is that it's high-risk behavior. In fact, a mountaineer who climbs recreationally, as I do, is about three times more likely to die from heart disease in the United States than from a mountaineering accident. But, of course, when there is a mountaineering accident, and a rescue, that short-term drama that we discussed here is what plays big, and not the many, many safe expeditions.

The other thing that we will talk about in the discussion is the Hubble decision. On January 16th [2004], Sean O'Keefe, the Administrator of NASA, Ed Weiler, and I went out to the Goddard Space Flight Center to announce to the Hubbard Space Telescope servicing team that Mr. O'Keefe had made the hard decision that we were not going to return to the Hubble Space Telescope for a fifth servicing mission with the Space Shuttle.

This hit me extremely hard. I am literally a "Hubble Hugger," as I think many of you know. I've had the privilege of visiting the Hubble Space Telescope twice. I'm a professional astronomer. I know Bob Parker is here somewhere, he's another astronomer astronaut, and I'm sure he can appreciate how tough this was. But Mr. O'Keefe looked at all the elements post-*Columbia*, and, in fact, our last mission was on *Columbia* up to the Hubble in March of 2002. And he looked at the recommendations of the Columbia Accident Investigation Board.

Hubble has a clock, an internal clock. And that clock is driven by gyroscopes and batteries. And sometime in the next two to three years, the gyroscopes that are on Hubble will wear out, and Hubble won't be able to do science anymore. Not too much longer after that, the batteries will run out of juice, their ability to charge and recharge, and at that point, the telescope will go cold and won't be able to be recovered.

Astronaut John M. Grunsfeld, positioned on a foot restraint on the end of Discovery's remote manipulator system (RMS), prepares to replace a radio transmitter in one of the Hubble Space Telescope's electronics bay. (NASA Image # STS103-713-048)

So we have to get to Hubble before the batteries die. And if you look at the recommendations of the Columbia Accident Investigation Board, and if you say we're going to satisfy every single recommendation before we go to flight and you say that we're not going to succumb to schedule pressure again, then when you look at the risk-to-benefit of using a Space Shuttle, you put yourself in a real box.

One of the boxes goes like the following: Imagine that we press forward with a Hubble servicing mission with the Shuttle. We have the crew trained, we have the big team trained, and we're on the pad. You know, maybe even we have liquid oxygen boiling off and the hissing and the moaning. And, in the launch count, we find out that something's not working right. A computer is down, a multiplexer isn't working, some communication link on the ground isn't working. Our flight rules would say, "Don't launch."

But whoever is in the hot seat that day will feel enormous schedule pressure to launch that mission anyway, because Hubble won't wait. We're all success-oriented, that's what we drive to. And Mr. O'Keefe didn't want to put any manager in that position.

Worse, when we go to the Hubble orbit, we launch due east. And, so, the only self-rescue capability we have—and that's another very important element in mountaineering or any outdoor adventure or going down in caves and, certainly, in the Antarctic—is limited to what you really have on the Shuttle. And, so, early in his analysis, he said if we're going to go to Hubble, we want to have a second Space Shuttle available on the pad so that you could launch within less than 30 days, which is probably the maximum you could keep a Shuttle crew going in orbit, in case of a *Columbia*-like accident. Well, imagine the enormous pressure if you had to execute that—of the second Shuttle to go rescue the crew.

Would we do it? Of course we would. If we put ourselves in that position, we would do everything we could to mount that rescue mission. And the same thing if the weather's not good, if something's wrong with that second Shuttle. And, I think, about half the time, there's some issue that delays us. We're getting better and better. I know two of my four missions have been delayed by a number of months. Many other missions have been delayed even as close to a few seconds prior to launch, when an engine will shut down for good reason, and we then recycle to two or three months later.

That's not acceptable if we're doing a rescue mission, even if it is a best-effort rescue mission. So I think the managers would feel that extreme schedule pressure that would put another crew at risk. So Mr. O'Keefe just felt that, as the top banana risk manager for the Agency, he didn't want to put us in that position.

That's a tough call. We all love Hubble. Hubble does extremely important science and is, perhaps, the most important scientific instrument ever created by humans. So this hit many of us hard, and it's that emotional side that makes risk-analysis and decision-making so hard. Someone said that the decisions we don't have control over are the ones that we worry the most about, and the ones we do have control over, we worry the least about. Well, this is one that I know Mr. O'Keefe has worried the most about.

And it really is compliance with all of the recommendations of the Columbia Accident Investigation Board and where we have raised the bar to make sure that we fly safely with the Shuttle—as safely as we can.

What we also heard in this conference [is] that the only limit is our imagination. Absolutely true. So the question you have to ask—and I did ask the Administrator—"Okay, if we can't go back with the Shuttle to service Hubble, how can we service Hubble?" I didn't quite put it that way, but I came back and said, "If we can service Hubble without the Shuttle, can we go forward with that?"

And I explained to him that it might be possible to use a robot to service Hubble. Now, keep in mind that I'm proposing something that puts me, as a spacewalking astronaut, out of business. But that's exactly what we want to do. We want to take routine operations of servicing—things that we can do with robots, things that we pioneered using humans that now robotics can do—and replace humans in hazardous situations.

EVA, Extravehicular Activity, is a very hazardous activity. We've been very fortunate in our spacewalks and there have been some close calls. That being said, servicing Hubble robotically will be a true, high-performance challenge. So, it's not clear that we can do it yet. But Mr. O'Keefe said we can go investigate that. This was an idea that came out of the extremely talented team at the Goddard Space Flight Center led by Frank Cepollina, one of our top inventors and out-of-the-box thinkers, a true explorer, one of the people responsible for the first servicing mission.

Remember, Hubble was a "space turkey," a "dog in space," "space junk." All of those things that we heard after it was launched, just because the mirror was ground to the wrong shape. Now, it was actually the best mirror ever created, but it was the wrong shape. Well, we went up and put contact lenses on it, corrective optics, on the first servicing mission. And for three years, people [had] said, "You can't do it. People can't service it. It won't work. It will be too hard. You'll end up destroying the telescope." But we did it.

It was that same team that came forward and said, "We think we might be able to send robots to the rescue." Well, that alone wasn't quite enough to put us collectively over the edge to suggest that we actually should proceed with the robotic mission, until we started listing the key technologies that we would have to prove to be able to service Hubble. And those technologies were: autonomous robotic rendezvous with a spacecraft, proximity operations close to the telescope, reaching out and grabbing the telescope, effectively a docking; doing an assembly, putting a new spacecraft underneath the Hubble—robotic assembly, and then, having dexterous robotics, agile robotics that can feel, to be able to service the telescope the way humans do, new tool development.

We looked at that list and I said, "Boy, that list looks exactly like our top list of things we need to learn how to do to explore—to go to the Moon, Mars, and beyond." And so the idea came up of using Hubble to be a catalyst for exploration.

Because, after all, what is Hubble? Hubble is out exploring the universe. It's our eyes for exploring the universe. It already is doing our exploration mission.

And what a great part of the Hubble story that it can continue to do science and be that spark that allows us to go further on.

Hubble is very hard to work on. This is going to be pushing what we've done in space, you know, maybe one or two generations. But people perform at their highest when we give them high-performance challenges.

If you ask people the easy things, they'll do it, but when you ask people who are passionate to do something hard, they'll do it well, and they'll pull out all the stops. And we've heard that in all of our panelists and in contributions from the audience.

Once we installed the Advanced Camera for Surveys in the telescope, there are a lot of very delicate operations. One of the good things about a robot is once its done the operation once, it can repeat it over and over again. And we have a greater knowledge of the metrology—of all of the measurements of Hubble—than any spacecraft ever, due to four servicing missions and all of the metrology we did on the ground—all of the measurements we made. So that we could build instruments on the ground while Hubble's in space and know that they'll fit. So this is the best setup we're probably going to have to try some of these hard things.

The proposed robotic servicing mission will launch on an expendable vehicle. We've got to get it up there sometime around late 2007 or 2008, so those of you who've heard about the MER, you know, 34 months wasn't long enough; well, we have about the same amount of time. It'll have two parts. It'll have a part that's going to stay with the telescope and one that will leave. Once it's on orbit, a robotic arm, much like the Space Shuttle, will be deployed. And there'll be people involved—this is not a push a button and it goes. Folks on the ground will be monitoring this and, maybe, controlling it.

We're going to grab Hubble in exactly the same way we did with the Shuttle. We'll have the same end effector—a very similar arm, a similar approach—and we'll use the same spots on the Hubble that we grab with the Shuttle. So, we're still in known territory, we're just using a robot. The robot will then put itself on the bottom. And that's exactly what we do with the Shuttle—we grab Hubble and we put the Shuttle underneath. And we latch with these exact same latches. Well, now the robot's doing it.

Once we let go, now we have to get the arms and the hands. To do that, we're going to use this special-purpose dexterous manipulator. It's already flight-ready. It's a Canadian arm called "Dextre"—that's the call sign. And it was built to service electronics on the International Space Station.

Well, we're going to steal that first and use it on Hubble. It'll deploy some cables to hook up the new spacecraft into the Hubble. This is something we feel comfortable doing, except for the part with the connectors. The connectors are always tricky—they're tricky for people with hands, especially when you're wearing these big space gloves. In two weeks, we built prototype tools that were able to take these connectors on the ground-based version on and off. And so we're reasonably confident we'll be able to do this.

But of course, we have to expect the unexpected, and Hubble always provides unexpected surprises. So we're going to involve all of the spacewalking astronauts and folks who have controlled robots on Hubble to go through all of these and think what could go wrong, and make sure we have the robot designed to do that.

Next, we're going to take out the wide-field camera. There's just two bolts and a ground strap and it pulls out externally. And we're going to stow the old one and put a new one in. The nice thing is, once we've taken it out, the robot now knows every motion it takes to put the new one back in. And we'll have extra cameras. One of the things about this robot is it can actually feel force. We'll have monitoring so that if it hits something, we see the force on that particular joint rising, and we can back off a little bit and change the attitude.

We also have to hook the new spacecraft into the brains of Hubble. That'll be another connector. It will be on the computer that Mike Foale put in. There's a connector on the top, fortunately, not on the side, that's just a shorting plug. So, take it off and put a cable on and then close the door on the cable.

Now we get into the really hard stuff, which is to take the corrective optics out. They are not needed anymore. All new instruments have the corrective optics built in. These are the sides of the refrigerator, so this is doing the job of Jim Newman, who had that advance camera and put in the Cosmic Origins Spectrograph. Once that's done, the servicing part goes away. Hubble, hopefully, will get between three and eight years of extended life, and then, at the end of life, we have to safely deorbit Hubble. Again, it's a safety and a risk issue. There is about a 1:250 chance that some large part of Hubble will survive to the ground in a populated area, and that risk is just too high.

So that little package on the bottom that has the new batteries in it also has deorbit engines. So Hubble will deploy all of its booms, start charging up the new batteries, do its science, and then sometime, perhaps as late as 2015, we'll feather the arrays, much like Mike Melvill feathered the wings on *SpaceShipOne*, and fire the deorbit module.

I'm hoping to be on a cruise ship somewhere in the Pacific to watch Hubble fly over and reenter. I think we'll all have to have a big party and really celebrate an incredible voyage. It will be this voyage that will have helped stimulate and advance us, probably by five or six years, in the exploration effort.

So that's the plan for the robotics. We actually have some contracts in place now. I think it was last Friday we announced that Lockheed Martin had won the contract to build part of the spacecraft. It is going to be assembled at the Goddard Space Flight Center as an in-house project. People say three years is impossible. But the really good news is that we have a tremendous amount of hardware already built, because we were on the road to a Servicing Mission 4.

Hubble has produced great images. How many people have seen the Hubble Deep Field? Or the Ultra Deep Field? An amazing picture. A thousand galaxies. It took 11 days staring into a blank part of space. If you hold a soda straw up to the sky at night and look through it, that's about the area of the picture. If you look at what's in the background in this relatively short exposure with this new camera, you see a lot of things. If you add up all of the spots, each of which is another galaxy, there are six thousand galaxies. Remember, that's just a little soda straw with a relatively short exposure. Each one of those galaxies has 100 to 200 billion stars. You heard me say yesterday, 10 years ago we didn't know about any other planets outside of our own solar system. Now we know, in just the nearby stars, of over 125. There are about twice that many that are being investigated to be confirmed. We now think planetary systems are common. So you can add it up just in this picture alone: 6 thousand times 200 billion times a couple of planets per star. There are a lot of planets out there. It's pretty mind-boggling.

That's where we're going on Hubble. Where are we going next with the Space Shuttle? Well, we're going back to the Space Station. In the President's vision for space exploration, our first task is to return the Shuttle safely to flight. The team is working through that. They are working through it with a passion as well. It is also hard.

> I HOPE TODAY IS THE START OF A DIALOGUE THAT YOU ALL HAVE WITH US AND THAT WE HAVE WITH YOU, . . .

We are finding a lot of challenges, not the least of which is we don't really have a Shuttle we can launch to test the new foam changes. So, we have our best and brightest engineers working on it. We have Admiral Cantrell helping us with the safety issues. We have really pulled out all the stops.

The crew that is going to go is led by Eileen Collins and piloted by Jim Kelly, both very experienced. I flew with Wendy Lawrence on my first mission. She's an incredibly hard worker, intense and talented, from the United States Navy. Charlie Camarda, Andy Thomas, Steve Robinson, Soichi Noguchi from JAXA, the Japanese Space Agency. A really exciting crew. We are going to dock with the Space Station. We are going to evaluate our techniques for inspecting the orbiter and for repairing the orbiter with some EVA flight tests.

The crew patch [referencing presentation slide] has the crew names around the outside and on the orbit. It has this swoosh and the STS-107 outline from their crew patch in recognition of that crew.

Before you leave today, we have STS-114 pins for every one of you. I want you to wear those as a reminder of this conference and the work we have ahead of us. There are going to be hundreds of risk decisions, reward considerations, and judgments that we are going to make before we return to flight.

I hope today is the start of a dialogue that you all have with us and that we have with you, and that you will take with you to the groups with whom you work, whether it's within NASA or outside of NASA. To continue this discussion, either as organizations or as smaller groups or as individuals, we can come back and keep this an ongoing dialogue. I think this continuing dialogue is very important if we are going to minimize the risk and maximize the return in our endeavors. STS-114 is part of that dialogue, and you will all be part of it now—those of you within NASA, by definition, and those outside—because of your participation here. We thank you for that.

As [NASA] Chief Scientist, I get to spend a little time in the House Science Committee room in Congress, probably more than I'd like. There are some things written on the wall that I think are really fantastic, and every time I sit there, thinking, What am I gonna say? or What are they gonna ask me?, I look up on the wall and read, "For I dipped into the Future, far as human eye could see; saw the vision of [new] worlds, and all the wonder that would be." That's from Tennyson. Again, this is something that I think drives us all.

Discussion

JOHN GRUNSFELD: With that, I would like to really open the floor completely to any discussion we might have, to talk about some of the overarching themes that came out of this meeting, anything we haven't covered. I want this to be very personal, and if folks don't ask questions, and even if they do, I'm going to ask folks that haven't participated yet to volunteer. In fact, let me try and stimulate that a little bit.

Why don't I ask our moderators if I can put them on the spot: Dave Halpern, Chris McKay, and our dinner speaker, Mike Foale, put you on the spot. We'll just start talking a little bit.

I think it was clear to me that there was general agreement that the greatest risk is not to explore at all. I think that is something that we have to communicate to folks. That we should not get so risk-averse that we just don't go out and explore. I have covered this one, but also the greatest risk is the lack of imagination. There are a lot of "greatest risks." They are all up there as the pinnacle of greatest risks. I had written down in my notes: Always expect the unexpected. Just like being asked to come down and talk. The other one, which I really liked, was Miles O'Brien saying, "The public is not as wimpy as we think." I can see no greater example of the public flying a new space ship than *SpaceShipOne* this morning.

Last year in the United States, about 40,000 people died in car accidents, of which 22,000 were not wearing seatbelts. I call that stupid. Folks who do that aren't thinking about the risk and consequence. In exploration, it is harder than that. Mike Melvill, on his first flight, and, as far as I know, on this flight, had no pressure suit. When you get above 50,000 feet, the remaining atmospheric pressure is such that the partial pressure of carbon dioxide and water vapor in your lungs dominates, and oxygen cannot get into your blood. Water at body temperature at low pressures will start spontaneously boiling if you expose it to a vacuum.

Mike Melvill had no pressure suit. That's hanging it out. Why? Why wasn't he wearing a seatbelt and a pressure suit? Well, performance. They are at the edge of the performance of what we can do with a vehicle like that, and that's the decision they made to take that risk. It paid off.

RISK AND EXPLORATION: EARTH, SEA AND THE STARS

OPENING PHOTO:
Commander Yuri Usachev (left) from Russia and U.S. astronauts Susan Helms and Jim Voss participate in Soyuz winter survival training in March 1998 near Star City, Russia. The three became members of the second crew to live aboard the International Space Station.
(NASA Image # 898-04118)

They have a very, very good safety program. They look at all of the considerations, and they consider their pressure vessel to be one that can't fail, at least given the risks that they are willing to take at this stage. So they are true pioneers.

In aviation, so many pilots died early on. As a result of that, we developed lots of the safety mechanisms that test pilots like Mike Melvill use to fly in *SpaceShipOne*. This risk-judgment-benefit is really tough stuff, and I think it's great when it succeeds. We need to keep pushing.

Miles also said, "Exploration is driven by fear, greed and curiosity." I like to stay on the curiosity end of things. Focus on the target. Get the right target. Have an overarching shared goal. It is just wonderful now that NASA has that in the vision for space exploration. We need to keep focusing on that. The other thing that I thought was interesting is that 96 percent of the energy content of the universe is in stuff that we have absolutely no clue what it is. I think that's great.

What I didn't know is 96 percent of the undersea environment has not been explored. Folks say the easy targets have been explored, and it gets much harder from here. Largely, that's true. Technology is no substitute for experience and leadership.

I am going to put Mike Foale on the spot right away, because I've talked a little bit about NOLS [National Outdoor Leadership School], and technology is no substitute for experience and leadership. Mike, in the Astronaut Corps now we are doing some things to try and enhance our expedition leadership. Maybe you can talk about just a couple of those things, like the NOLS, like the NEEMO [NASA Extreme Environment Mission Operations], and let folks know what we're doing.

MICHAEL FOALE: Well, John, I covered a bit of that in my dinner talk. To carry out any technologically advanced exploration mission which involves complicated techniques and new equipment, you have to be trained. You have to know how to use your equipment. If you take really fancy gear with you and you don't know how to use it and you waste time on it in a blizzard or on top of a mountain, you may actually end up risking more by messing with it than leaving it behind.
When we train, we are basically putting aside that risk. We are mitigating the risk of carrying out the mission when we actually, finally, get to space. With astronaut training, just as the training we heard about going under the sea or into the deep caves, certainly you train for mountain climbing, we try and train so that we will perform better once we are there.

However, the training is not necessarily without risk. The training is not necessarily in the simulator, where I think you're pretty safe, unless a brown recluse bites you, which happened with Joe Engle once. Some of our training involves going outdoors. Some of it involves flying in aircraft that could have a malfunction, in particular, as we look toward exploration beyond Earth orbit, and in particular, when we can't turn around. Think of the Apollo 13 example, when your problem occurs on the way out to your destination, or where you are forced to go a long way out before you get to come back, then you are in a situation of survival or making the best of all the materials and resources at hand.

DISCUSSION

We have started to develop training that will take astronauts in small groups, crew size roughly. Six is what we've been choosing. We are then working with the National Outdoor Leadership School for one, the Canadian government for another, and the NEEMO group in Key Largo, Florida, to explore and develop leadership, followership, and self-management skills in our crews. That is one task.

Second, we are starting to get a feel for what we need to do when we are isolated and when we are dependent only on each other and the things we have at hand. That training we are still in the process of discussing and inventing.

I showed you three cases. One is the National Outdoor Leadership School, where we go out either into the Canyon lands or into a mountain environment. Another one has been exercises. We have taken part actually starting off in Cold Lake, Canada, where basically we are given a scenario more of "expeditioning" from one point to another. It is kind of a fancy series of walking through the snow, managing yourselves, looking after your team, and managing the equipment that you have with you. It is led by instructors who already have a plan on what the exercise is going to be, and it is covered in terms of risk, because we also have the full resources of the Canadian Armed Forces to get us out of there if we actually got seriously hurt.

Another analogue that we have been exploring and using is the Aquarius Laboratory in Key Largo, Florida. There we've put together crews that are three astronauts with three nonastronauts. We have actually included our Mission Control flight leads, who would normally be in the Control Center controlling a mission, to take part in those dives and those missions. And there we have actually solved another issue which is the classic problem of "What on Earth is ground control thinking!" when you get these strange instructions. Especially if you actually have people who have a stake in your activities. Scientists who are not in the team but [are] back at home in safe conditions will be asking things of you that might be rather difficult or seem rather strange or irrelevant at that point in your difficulties. And we're trying to bring together the Mission Control teams with the astronaut teams that would be deployed so that they would see each other's problems. And we've done that two or three times now.

Looking to the future, as we plan for moving to the Moon first and then Mars, we need to develop further the idea of being able to maintain our equipment, look after those resources that we have, even if they break and are a long, long time away from any kind of refurbishment back on Earth. To that extent, we want to actually follow-up on some of the Apollo lessons that we saw to teach geology at first. And these were in the deserts, I think always in the United States. Because of the Martian meteorite interest now, it's very exciting. We have thought about attempting expeditions, taking part with other scientific expeditions where there is a scientific goal that we, the astronaut office, do not have a stake in but they have a stake in our performance. So, that would be realistic and an analogue to a real mission on Mars or the Moon, where we have to carry out some of the grunt work—the deployed work—that would be required.

We are also looking at going to hot deserts, where meteorites also occur. This would combine the expedition training with acquiring field geology skills that would also be required of a small group going to the Moon or Mars. I think that summarizes it.

I should just add, once we get to these deployed situations, there are risks. And I myself was a little perplexed a number of years ago, when I was asked to come up with the whole safety/risk mitigation plan. And I had to have a safety review of what we were talking about and planning. At that time, we were sending roughly six astronauts at a time for survival training in Russia. And at that time, I talked about needing to have some insight into the other partner's processes. The same would go for the Canadians. The same would go if we hire another group, such as NOLS, for example. NASA has to know what you do when we use your services.

With Russia, they had had a number of helicopter accidents, and the training we were proposing was to do some of those search and rescue exercises with the Russian helicopters. And in the end, people were so just alarmed by the stories coming out of Russia we had to turn that off. I didn't know how to fill out a kind of safety/risk plan or matrix on that. Sometimes, you just have to use good judgment. You use your intuition. You ask all the questions that you can. We heard how James Cameron manages his film set. That was interesting to me, because he doesn't have the formal process that NASA has in its bureaucracy, forced on as a result of many, many mishaps over many, many years of experience. Sometimes, in smaller groups, you have to use judgment. And I think we are going to be in a position of having to use our judgment as we assess some of these new activities, not only just processes and safety reviews.

JOHN GRUNSFELD: One of the things that you brought up was working with our international partners. That, I'm sure, is going to be a major issue with pushing out beyond lower Earth orbit to the Moon, Mars, and beyond. And it happens on Earth frequently, the issue of different international cultures. And we heard a little bit of that. I think that's one we're really going to have to grapple with. Not just the day-to-day living types of things, but how different cultures deal with the safety issue. Whether it's documented or whether it's trusting people's good judgment.

MIKE FOALE: I know Chris, and certainly others from Ames who have been doing research, they've had to work, in particular, with Russians. And I know James Cameron did. I'd love to know what their opinions are and how they manage insight into systems that they don't know all about.

CHRIS MCKAY: It's difficult, and in Russia, in particular helicopters flights—flying with the U.S. Navy in Antarctica is very different than flying with Aeroflot in Siberia. And we'd like to take you on both trips, Mike. Actually, we would like to get you into the ice-covered lakes. I think you and Dale and I ought to talk after. But, actually, what I would like to do is come back to this conference and just think about it a little bit. It's been an incredibly fun and interesting conference. I can't remember when I've enjoyed one as much. My question to the audience is, how do we make it a useful conference? How do we take what we stated here

DISCUSSION

and make it something that could be useful—and I would say, useful to whom? Useful to three parts of NASA. The robotics program, which, after we heard Steve Squyres talk, I think is clearly in need of a better risk assessment strategy. The near-term human program, which we're hearing about. I think John, Mike, and others are well involved in that, and I think that's been in the impetus for this conference. And I applaud them for their efforts. And I think that's the light that makes a way of doing something useful possible, [your] attention to this. And, then, third is, [as] Mike was just saying, long-term human exploration of distant planets—which is going to be a completely different category of risk and danger. So, how do we make this conference useful to the robotic program, to the near-term human program, and to the long-term exploration program?

And I know that there are a lot of people in the audience with a lot of good ideas about this, because I would hear them, as I was scarfing down my dessert at the dinners and lunches. And I think it would be good to get a dialogue going. How do we make this conference useful, rather than just all going home and having had a fun and interesting time?

QUESTION: Andy Presby. I'm student here at the school. I'm glad you asked . . . The sign on the wall there behind you says, why do we explore? And I think I've heard a lot of very inspiring and interesting stories over the last couple of days about why individual people—panelists and people in the room—have chosen to explore. And a lot of them are the same reasons that NASA has inspired me pretty much since I was born, since I can remember. But I think an important thing for you guys to realize is that the first thing that struck me is [that] not all of you explore for the same reasons. And when you're looking at NASA from the outside in—and I think some of the folks from Hollywood and the media have identified it correctly—the public is not as concerned about risk maybe as the explorers are. The public seems to have sort of understood that you guys accept the risk and you do it because you love it, for whatever reasons.

What the public is worried about is, why are we going? And why should I pay for what you love to do? Why is it helpful to me to pay for what you love to do? And I think that if you guys walk out of here with anything, perhaps a useful thing would be an internal dialogue amongst yourself culminating in an intensified outreach program to explain to the public, in terms that they can understand clearly, why they should pay for what we all in this room, I think, would agree is one of the most important things our government does for us in this country.

MICHAEL FOALE: I think we heard very eloquent expressions over the last three days as to why we explore. I'm actually more worried about the public not perceiving when it's dangerous. I don't believe people expected the *Columbia* accident. Astronauts do expect the *Columbia* accident. And I think there are misconceptions out there. Someone referred to it. It's the repetition of anything that makes us numb to the risks. And because we've seen Space Shuttles launch and land successfully a number of times, it was a surprise.

The risk has not been well communicated. What *SpaceShipOne* did today was extraordinary. And you saw, if you were watching, how something very unexpected happened during the ascent. The Space Shuttle goes—there's 25 times more energy in that whole business. I mean it's 100 tons, is it? Take the speed, divide by 25, and square it, you're going to get the answer.

JOHN GRUNSFELD: It's 25 squared over 3 squared. [laughter] We're two physicists. We'll have this in a moment.

MICHAEL FOALE: So if it's Mach 3, and we go Mach 25 in a Space Shuttle, divide 3 by 25 and you get about 8. And then you square it, and it's 64. But it's huge. The difference is that their heating on entry is just going above boiling—if that. It's not anywhere near risking a metal hull. If it's a composite hull, it's going to start risking it pretty soon.

A space vehicle gets up to 2,000–3,000 degrees Fahrenheit. So these issues are engineering issues, they're mundane, they're arcane to the public, who don't really care to hear the details. But the final answer is that it's dangerous. It's risky if any of these things fail.

John brought up an interesting comment about the risk of this launch today, which I would like to get to, to tell you that there is risk here even in *SpaceShipOne*. He talked about the lack of a pressure suit. He talked about the need for closing the hatch and living only in shirtsleeves there. We don't do much different on the Space Shuttle. We have pressure suits, and we have parachutes. They didn't do the *Columbia* crew any good. I don't know they would have done the *Challenger* crew any good.

So the situation really isn't so different. And, yet, John pointed to the risk this morning for Mike Melvill as he did that climb. The risk is still there for every Shuttle astronaut that will be flying on the Space Shuttle henceforth.

JOHN GRUNSFELD: I was trying to use the seatbelt analogy. It doesn't guarantee it. But it reduces the risk. Good point.

CHRIS MCKAY: Why does the public think then, that NASA is going to make it risk free? There is the perception that if we were disciplined, if we followed the *Columbia* accident report rules, and if we had a culture of safety, we would be risk free. Somehow the message that you guys are saying, which is that it is inherently risky, people are going to die, crashes are going to occur, is not being conveyed by NASA. We're not getting across the message that you're articulating.

And that's what I'm saying is, how do we turn this conference into something useful? Well, maybe we need to start figuring out how to get that message across, and stop giving the impression that we can make perfect systems.

QUESTION: Tom Krause, BST. We're involved in assisting with the culture change effort at NASA. It seems to me that the issue is not so much that the public doesn't recognize the risk, but rather that the public finds unacceptable the possibility that something could have gone wrong organizationally that led to the accident. So, when the investigation finds that errors and mistakes were made that could have been prevented, then it seems to me the public says, something about this just isn't right.

CHRIS MCKAY: Can I react to that for just a bit? If you take just about any accident and trace it back, you can find a step or a place where it could have been prevented. That's just the nature of these complex systems. And I don't think that you'll ever be able to come up with an institution, a large group or even a small group, where your accidents due to human factors or human error are gone. I think that's unrealistic. And maybe Scott, who is on the Board, we might put him on the spot here, since I think he works at the same Center I do, could comment on it.

SCOTT HUBBARD: Yeah, let me see if I can parse a little bit from where you're coming from, and what he's saying. It took a long time to be sure we had the physical part [about the *Columbia* accident], in the end we got that with no equivocation. Everybody absolutely knew that. The organizational part took a lot longer, or took a different approach, and was in many ways more complex to understand. And, I think, having people come in and talk to us, having members of the community as well as experts in behavior and complex systems and human factors talk to us, the distinction was that we had, perhaps, led the public to believe that we had done everything we could reasonably do. And, in fact, as we peel the onion on the accident, we found that there were cases where, because of repetition of something that started off as an inflight anomaly and became a turn around issue, because of other situations where people had fallen into poor habits of engineering analysis and so forth, we really did make some human errors that, with a different type of approach to it possibly, could be addressed. And, in fact, that's the result of the culture change.

So, now what we have to do, I think, is to tell the public that there is a level of risk. That we are doing everything we can to mitigate that risk, but it is not going to go away past a certain point, there aren't perfect systems. We are going to address the culture issues as much as possible, but there is going to be an irreducible residue in there that you're going to have to deal with. I think the danger is that, with the talented people in this room, and the Astronaut Corps in particular, you make it look so easy. All the thousands of people that support, with all the things that are done, the impression comes across—whether it's in the robotic program, with the perfect landings of Spirit and Opportunity, or whether it's with the Shuttle program with, by all accounts, a perfect takeoff and landing—that we've got it down.

The fact is that anybody who has participated in a launch, particularly if you've been in the position of being the last person to say go, and you hear in the background, through your earphones, all the thousands of things that have to be right, all the systems that have to be polled, you know that there is an irreducible risk of something catastrophic happening.

We do not tell the public that story. I think if the public just had the earphones on of the guy in the polling chain as you're getting ready to launch and was aware of all the people at all of the systems and all of the things that have to happen, they would immediately realize, gee, what we have to go through to make this happen, it's truly extraordinary. So, I think that that's part of what we need to communicate, and part of what this business is all about.

JOHN GRUNSFELD: Let me put John Gans on the spot. He's the director of the National Outdoor Leadership School. They have thousands of students every year who go out into the wilderness, go out into risky situations. I imagine occasionally a parent will call and say, my son or daughter is going to go out and do this rock climbing, is it safe? And how do you communicate to them the risk element as an institutional risk manager?

JOHN GANS: Well, first off, we try and be as clear as possible that we can't guarantee anyone's safety, and we're up front about that. I think every time I get on United Airlines and I hear, you know, "Safety is our number one priority," it runs through my mind that, no, getting us there is the number one priority. Safety may be number two. But say safety is number one, we wouldn't take off.

JOHN GRUNSFELD: It's clear that profitability is not number one.

JOHN GANS: So we try and be as clear as possible. And you mentioned the parent-child thing. I'm going to switch the question some, because Dr. Sylvia Earle talked about the role of education as it relates to exploration and risk management and getting people outside and other things. And I think the interesting thing that I have been thinking about in this conference is that my daughter, this summer, started climbing in a more aggressive way. Safely, but in a more aggressive way. My daughter was 10 this summer, and she wanted to get ready to climb Devil's Tower with me this fall.

I adore my daughter, you can probably tell. And it really hit me that, suddenly, I'm on the other end of this, and I'm hesitant about what she was going to do. Now, climbing has been one of my passions in life. That's where I've felt most alive. It's where I've had some of my best relationships with people around me, with the world around me, and the environment around me. And, suddenly, was it okay for my daughter to do it when it moved beyond the walk-up situation into something that was more serious? And I came to terms with it. We are going up to Devil's Tower in October.

But there is something about generational passing as it relates to risk management. And we certainly run into it with parents making decisions for their children. It certainly is tied into the educational issue. But it's something that goes to each individual family, and it's something that I've thought a lot about over the last few days. It's something that goes to the space program, the generational difference between the people that grew up with Apollo, the generational difference now. Look at the number of parents now that won't let kids go off and ride a bike alone, wander out of their neighborhood alone, whatever else is the case.

I realize I'm broadening the issue far beyond NASA here, but it goes so far beyond what we're talking about here, and, somehow, I think there is a role to play for our society in making the parent-child relationship understand risk better. And there is a role there for our schools. So, now that I blew up your question totally into something else, I'll pass off the mike and not go on further. But the long and short of it is, we try and be very clear with parents that we can't guarantee any safety out there, but we manage it very well, and then we convey the benefits. And we know the benefits right down the list, and we rock at all those benefits, and are clear about them, much the same conversation that's gone on here.

JOHN GRUNSFELD: That's great, and it brings up another point, that you brought out, which is, if you think about the early part of the space program, prior to the first American going into space, rockets generally blew up. Most rockets blew up while we were trying this.

UNIDENTIFIED SPEAKER: They still do, John.

JOHN GRUNSFELD: They still do. Not most. Some. Let me just take this a little further, which is, when Michael and I were growing up, that's what we saw. We saw the struggles, there was no question that it was risky. And as we started flying more, and then we built a spaceship that looks like an airplane, it brought it into everybody's daily experience. Then, people who are growing up now, like my children, space is part of their culture. It's become the norm. And so people don't really notice the space program now until we don't have one. And I think that's an indication that it is part of our culture, and that the education can help. Go on, Mike.

MICHAEL FOALE: I just want to add that people in this room are probably aware that . . . I don't know of any rocket system that can launch 1,000 times and not have an accident. Most rocket systems launch 100 times and have an accident. So if that is the only way, if you're on the rocket on the 100th time, and you do a lot of trials, and you do the statistics, that's [it], you die on that rocket. So the way you get better than one in a hundred on any rocket system is to have a way of surviving that explosion that 100th time. And the Russians have done quite a good job with the Soyuz escape system, it's worked twice in all of their launches, hundreds of launches. Apollo was a good system never used. I think Gemini has an interesting case. It's a story as to why they didn't have an escape system quite like the Mercury before it.

But that is the way we get away from those—the fact today is that rockets do still blow up, and we can't do anything about it right now. We don't have a strange, wonderful, anti-gravity technology that will get us away from that.

JOHN GRUNSFELD: And one in a hundred is the best of the best. Most are not nearly that good.

QUESTION: Joe Fuller. I'm sitting here very anxious, because I don't think we're getting down to business. It's been a wonderful conference, you know, over the last two and a half days, and I think we've learned a lot. The problem is, how do we capture the knowledge that's been just flowing out here?

At some level there's a connection with the way we do business, and we need to search for that. So what I would suggest is that, the first thing, we capture the proceedings of this, and the second thing is, we form some kind of organization, you know, ad hoc or whatever, to pursue this information and make the connections that are so obviously there.

[In] some kind of way, the institutions have got to get involved in this. I think that, as someone said, every individual has a value calculation that they have to make, and they have to make that trade. We can't determine the perception of risks for the individuals. We can't determine the value for them. But what we have to do, and what we do in business is, the value proposition has got to be so large that the risk is acceptable.

So what I would suggest is that you've got to go farther than this. You can't stop here today. You've got to put some organization in place to carry this forward, and mine this knowledge for the value and the benefits that are obviously inherent in it.

I'm involved in risk management professionally. I haven't seen too many other people here that are. I did hear Mike Gernhardt talk about how he's using quantitative risk analysis. So I would volunteer to be a part of that group, to determine a strategy for extracting the knowledge and information so that it would be more useful and of value as we go forward and explore.

CHRIS MCKAY: I have a suggestion. I think that's a good suggestion, how do we connect to the institution of NASA, in particular, the results of this conference? It seems to me [that] to do that you need someone who is close to the Administrator. He clearly wants to get advice on this topic. Someone who is passionate with experience in this area. Somebody like the Chief Scientist, John Grunsfeld. I think we should add to his responsibilities this area. I think this would be a perfect opportunity. You've seen the conference. You were obviously one of the ones who put it together and organized it. I really think that the mantle falls on you to carry this forward within NASA as an institution, not just the near-term flight program, the return to flight. But also thinking long down the road.

Also, I think the robotic program is in need of a clear-headed assessment of risk. Now there the risks aren't to lives, but they are to resources. And I think that that program also needs a clear risk assessment. And I think the Office of the Chief Scientist right now is a good place to do it. So, all voting for John as the representative of this?

JOHN GRUNSFELD: Thank you, Chris, for your kind comment. But, seriously, I think we have Tom Krause here from BST working on our culture. This is something that the Office of the Chief Engineer, that my office, the Office of the Chief Scientist, Bill Readdy, Office of Space Operations, Space Operations Mission Directorate—this is a dialogue we have everyday. And we wanted to broaden that from NASA management to you folks, and, as I said up front, the start of a dialogue.

But the other point was capturing this and you were just captured. You were captured on videotape. We're going to convert that. We've been talking, prior to

DISCUSSION

the conference, about how are we going to put all this together. A number of folks have been chronicling this individually, but we're going to do it institutionally as well. And I know, Keith, you've thought about that a little bit. Do you want to say anything? Let me put Keith on the spot, and then back to you.

KEITH COWING: I think we are quadruply redundant here. I am recording this on my iPod, so I can be listening to it as the transcripts arrive in my e-mail box in about an hour up at Ames. We hope to have this online in a very short period of time, just the raw verbatim transcripts, with the "ums" and the "ahs" and the spelling errors taken care of. I'd have to talk to Bob Jacobs and some of the Ames folks to get the specifics on what the follow-ons are, but there is talk of putting some of this on a DVD, of putting a more comprehensive document together. Steve Dick and I have talked about something more comprehensive, in terms of a history monograph. So the initial concept here, John saw the first e-mail that started this, was capture everything in as many ways as possible, so Joe, you're psychic, you knew what we were doing when we were first doing it.

QUESTION: Scott McGinnis: I'm a student here. What we do in the military, and I'm sure a lot of you are military, but if you assume this conference is like a six-month deployment, every time before you leave the ship you have to give your lessons learned. You get one line, everybody. Then the XO, I think, Dr. Grunsfeld, that's you in this case, forces everyone to read it prior to going on [to] their next deployment.

So that built a database, and as the XO you are required to make sure that they all sign and verify that they have done this, proving that they have read it. Then, when they make the mistake again, the responsibility then lies with the responsible individual, the person making the mistake. Therefore, you have a traceability and a responsibility for each individual action, and also, it shortens the amount of data; instead of having to watch our three days of deliberation, being able to shorten that and pull out the small pieces.

So that's part of the military structure, and you've got a little more discipline—I think we talked about the flogging and all that kind of stuff. [Laughter] We have a little more coercive nature in the military to be able to do that. But I'm sure NASA can muster that up. And second, you're talking about the [pressure] suit of *SpaceShipOne*, the risk that they're taking. And I think it all goes back to the benefit that we haven't discussed, we've tapped around it. Dr. Spudis brought up the three reasons why we explore.

I think the fourth, and Magellan showed it with his cloves, is money. And *SpaceShipOne* is doing it, one, to explore. But come on, we've got a $4.5 billion market in the tourist industry. And they are exploring not because they want to prove science or prove humanity. We've proved we can do it with the money. But can we make it profitable? And I think if you saw the big "Virgin" on the side, and you saw the big Sprite advertisement going on, and the M&Ms floating around, I think we have found one of the keys to space exploration, and that is the good old American greenback.

I think NASA also needs to find a way to maybe encourage that, like we did in the early '20s with the prizes. I think we've tapped around cost as a benefit, or money as a benefit. I know NASA can't get benefit monetarily that way. But it's definitely a point we haven't brought up.

JOHN GRUNSFELD: No, we'll be offering prizes. We've got Congressional authority to do that for similar challenges. But it's clear that for Rutan, this is about his passion, about pushing new envelopes. The X prize is $10 million [he blew off] quite a bit more than that. And I think it's great that he's been able to leverage the commercial sponsorship there to help offset his cost of developing this Because it is opening a new frontier.

Eventually, folks who want to actually sell services will have to start incorporating more of the safety rigor. You probably wouldn't go out on a cruise ship today if you knew that one out of every four or five times you weren't going to come back. So that's, again, that comes down to the profitability. And what Mike Gernhardt said is, you have to have a successful dive operation to have a commercial operation be successful. And so, safety is a critical part of that greenback.

QUESTION: David Liskowsky from NASA Headquarters. I'd like to perhaps comment on some of the discussion that's been going on. We're at a point in time at the Agency where we've just gone through a large transformation to hopefully meet the exploration vision. I think we're all behind that, and that's what we'll be going forward with.

Maybe we can take this opportunity at this time to use these changes that are going on in the Agency to change our message. Change our message to the decision-makers, mainly Congress and the public, about what the nature of this business is, that it is risky business. Everyone talks about that, that NASA has been a victim of its own success.

But maybe it's time that, as we go forward with this new exploration vision, and this is something that can be done through John as Chief Scientist, we have the PAO [Public Affairs Office] folks who shape the NASA message let folks know, truly, what the nature of the business is, and to let them know that, as we go forward with this new vision, it is going to be risky. And without abdicating our responsibilities to meet the requirements of the CAIB [Columbia Accident Investigation Board] report, there is going to be that element of risk. And it's part of us shaping the Agency's message and how we convey it to the public.

Maybe this symposium can be the first step in trying to do that, in shaping what that new message should be for the exploration vision, and making, perhaps, a little more realistic vision than the Agency has had in the past.

JOHN GRUNSFELD: Good comment. Well, last comment, and we'll go to "Moose" Cobb—Robert Cobb—he's been dying to say something. Then we'll stay around for comments afterwards. In the packet, we gave everybody a pad of paper and a pen. And so, before you're allowed to leave, you'll have to write down at least one lesson learned, and provide that. You don't have to have a name on

DISCUSSION

there, but if you have any ideas, sketch that out and put it in the bin before you take your STS-114 pin, which I really want you all to have.

ROBERT COBB: I'm the NASA Inspector General. I've been with the Agency for two and a half years, with no exploration or science background coming into the Agency. It's my perception that NASA works hard to dispel the notion that what it does is inherently risky, and the reason it does that is because there is a fear that the public won't fund it if NASA tells the truth about the risk.

That's something that I think that this conference goes a long way towards—I think people recognize that the public is willing to accept risk. And that the idea is, the object is, that it's important for NASA to have a transparency into the risks that it is accepting and to allow the public to share in understanding of those risks.

QUESTION: I'm Sandra Cauffman. I'm from Goddard. I think we're missing some basic thing here. The question is, why we explore, and we are not really answering that question. We're talking about the risks and, yes, that is very important, but the people out there need to understand why it is that we're doing what we're doing and what they are getting in return. They like to understand why we are risking the people, but what are they getting back?

In the DOD world they understand why we are risking our soldiers and why we are sending people to war and whatever, but in the NASA world they do not understand why are we sending astronauts. And they see pretty pictures of the stars and stuff, but what is it that they are getting back in return as taxpayers? And we need to really send a clear message to them. And it's not PAO [Public Affairs Office] stuff. It depends on each and every one of us to do that.

Just a little story. I was in National Night Out in my neighborhood a couple of years ago and I was talking to my neighbor, a nice little old lady. And I am the Deputy Project Manager for the GOES-R Satellite. And she was asking me what I did for a living and I told her about the weather satellites and all this and all that. And she just looked at me with this puzzled look on her face and she said, "Why do we need weather satellites when we have the Weather Channel?" You know, that's what we have to deal with, the perceptions out there. Yes, the risks are there, but they need to understand, okay, we are risking, but what are we getting back? So, I just wanted to say that.

JOHN GRUNSFELD: That's a great comment. Natalie, why don't you take that?

NATHALIE CABROL: Actually, I would like to add on that comment because this is probably translating better, what these guys were saying yesterday. What is the gold? Not the goal, but the gold, you know? Five hundred years ago, Magellan leaves, and he brings back cloves and he brings back riches. What are the riches that we can show to the people today? And there are many. And we are good at it at NASA, but we are not good at telling people. You know, from the Moonwalks people today are going to ski better. They have good Moon boots, medication, things that we do in space better the health of people, the expeditions in the sky, in the sea, or on the land are bringing [generating important discoveries]. We have that, but we are not translating enough to the public. And I think this is where we need an effort.

JOHN GRUNSFELD: Steve?

STEVE DICK: Following up on that—I'm Steve Dick, the NASA historian. On Friday we're launching a series of essays on the nasa.gov Web site called "Why We Explore." And I think this will address some of the questions just raised. And it's not Public Affairs, it's historically nuanced and historically based. (And, by the way, this is the 46th anniversary of NASA, on Friday, October 1st.) The first essay will deal with why we explore in the sense that exploration is necessary for a creative society. And I'll talk about Ming China, which was mentioned by Jack Stuster the other day. That's on the NASA Web site at http://www.nasa.gov/ missions/solarsystem/explore main.html. And it'll be a once-a-month, "Why We Explore" series, a different essay each month.

NATHALIE CABROL: I will wrap up quickly. But, you know, why do we explore? I think within us it's just because we think that somewhere on the other side of the hills, as you were putting it, it must be better or something is better than what we have now. Otherwise, we wouldn't be doing it. And it's true that maybe the other side of the hill has nothing particular, but what we learn along the way is bringing a lot of good to society, et cetera. So we need to emphasize this really, really hard.

JOHN GRUNSFELD: And again, it comes down to both personal and institutional, as well as national. In the President's vision [for space exploration] he said, "The purpose of this is to advance U.S. scientific, economic, and security interests." And it's through a broad range of things. As you say, along the journey you learn a lot of things that improve our life here on planet Earth.

But it's also the higher purpose. You know, we're trying to understand where we came from. Why is there a universe? And in the process of very basic research like that is where we learn the really valuable things—like quantum mechanics that leads to lasers—that it would be a long time before you'd do that with just subsistence farming. So, these types of things are very important.

I also have something that often ends up resulting in controversial discussions, but I have a statement that I think is true. I can't prove it, but it's "Single-planet species don't survive."

QUESTION: Dave Leckrone, Hubble Space Telescope and NASA Engineering and Safety Center. I guess we're all ganging up on you because several of us must have made the same comment to you. So, I want to start out by thanking you for stimulating this conference, which has been absolutely fascinating. What fascinated me most in hearing all the speakers and the discussion and seeing the film last night about Ernest Shackleton and the *Endurance* expedition was this business of what compels us to explore and take these risks in the first place, instead of just adopting the fetal position in our lives.

And I have my own ideas. I actually wrote it all down and I'm going to exchange this for my pin later. But it sort of goes to what was said just a moment ago, and I think Scott Hubbard mentioned this on the first day. We explore because we have no choice. It's an evolutionary imperative. Our species became what it became because it explored. What was over the next hill was either a threat or a source of sustenance. And if there wasn't anything there, then you had to go to the next hill yet to check that one out. And I think this is built into our DNA.

Poor Ernest Shackleton was so obsessed with exploring he couldn't even really articulate why he kept going back to the Antarctic. He just had to do it. And I think at least some of us, if not all of us, within the species have it built into our DNA. And I think corollaries to this are all having to do with survival—acquisition of knowledge, commerce, education, creating a national identity, finding not only individual self-fulfillment but group fulfillment. And I think every one of those relates, going way back perhaps, to our need to survive as a species. And maybe we can't survive as a one-planet species.

JOHN GRUNSFELD: I agree with that absolutely. You know, we try and raise it to a higher plane but, ultimately, it is, I believe, hardwired into us to do this. But, as well, our evolution has taken us to be a species which is a thinking species, sometimes rational species. And, so, it's also provided us the ability to question what we do. And that's where this becomes a little bit messy, because we say, "Well, is it worth the risk?" And that comes back to where we are.

And if anybody doubts that we have a survival imperative to explore, just look at the situation we're in with science, technology, engineering, and mathematics in this country and where that may lead to eventually—because technology is the key to economic prosperity, which is the key to security, which is the key to freedom. And I believe that exploration is linked to our ability to stimulate people to, directly and indirectly, get a good education and make use of that productively.

UNIDENTIFIED SPEAKER: I'm just going to build on some of these other things that people have been commenting on. And, in particular, I want to play devil's advocate to some of the spinoff comments that have been made.

I agree that this is very important, and some of the discoveries have been fantastic. But really—and this builds on your comment earlier—I think that there's one question that NASA needs to be accountable to, or one big question, and that's quite simply, are we pushing the frontier? Are we pushing the frontiers of science, technology, and exploration in a way that no one else can—no individual, no company, no university—in a way that only NASA can? And that's the thing that we constantly have to be asking ourselves. And I think this conference is part of getting at that issue.

JIM GARVIN: Well, thanks, John. I think there's one comment notwithstanding the spinoffs and everything. I mean, we can all play the game as, Dave, you said so well about, this is an investment choice. It's part of our DNA. But I think it also bears witness to trying to generate metrics and look at what the impacts have been. And we do that perhaps ineffectively, as you've said, Nathalie. But they are not transparent, they are major. And if you ask some of the technology leaders, exploration has begotten these catalytic effects.

So without it, the question that you raised, John, is the one I think this group needs to raise. How fast would we have progressed in different areas? I mean, maybe Darwinian progression—you know, seeking optimization whenever we can—is not the game afoot, and natural selection in technology doesn't work. I don't know. That's a great thing to debate. You know, maybe Steve Dick's group in history can study that.

But I'm still struck by questions that when we ask people in other sectors of society—IT [information technology] being a good example—in remote sensing of this planet, the benefits, while maybe not tangible in terms of dollars in your pocket, are there. We would not have microcomputing with fault tolerance, ever. There would have been no imperative, except perhaps a very narrowly-defined security interest area—which is important, of course—without this exploration imperative. And we demonstrated that.

So I think we need to do better at defining those metrics. I mean, yes, the textbook metric, I think, is an important one that most people seem to forget. I like to think that all the textbooks have been rewritten in the last 20 years in many of the areas of astronomy, physics, planetary science, and even this place of our own planet.

But anyway, I think that's the amplifier on technology progress in areas that aren't the ones that have instant economic gain. That's what we should be doing, and that follows on what you said so well. That's NASA's unique role as a government agency. Otherwise, it would be private. Thanks.

QUESTION: I'm Becky Ramsey, NASA Headquarters. Recently we had someone do a study for us. And while it was a very interesting study, I won't go into the whole thing. But one of the stats that struck me is that a majority of the people we talked to said that they like NASA. They don't have a clue what we're doing, but they like us. And I think we cannot lose sight of the fact that we're not the only ones who want to go. It's not confined to the people in this room or the people who attended this conference.

I walked over to the little lobby bar last night. I was sitting there watching the baseball game, and I got into a conversation with the bartender and some of the servers. They said, "Are you with the NASA group?" "Yeah." "That's so cool!" You know, they don't know what we do, but they like us! And we have to build on that personal connection. We are their representatives. Until Burt Rutan starts charging five bucks for a trip into space, most of the people out there are not going to get to go. We have a responsibility to be their representatives and to do what they can't do yet. I mean, we talk about the spinoffs. They don't really care about the spinoffs. Yes, they're important. Yes, the benefits that we [generate] make everybody's lives better. But they don't know about that, you know? We tell them, but they don't read our cool little magazine. They don't know the weather satellites from the Weather Channel. They don't care that much about that. They like it because it's cool, because they want to go. And I think we can't lose sight of the fact that that's why exploration is important to everyone else.

JOHN GRUNSFELD: I absolutely agree. In fact, in other studies we've found that the NASA logo—the meatball—is likely the number one brand recognition. There may be a couple others that are close. The other thing we found out is that, when we were working on our renewed vision of discovery, we found out that most people assumed we were already doing all these things. You know, when we'd say, "Well, what do you think about having a renewed trip to go beyond low Earth orbit to the Moon and Mars?" folks would say, "Well, isn't that what you're doing?" And we'd say, "Yes, that's what we're doing!" And we have to communicate that a lot better.

MEL AVERNER: That's not true. We're not going to Mars and Moon. We are attempting to do that, but it's not our mission yet. And if we say, "Yes, we're going," people will go away saying, "Great! Great! You're going!" Okay, you got my drift.

JOHN GRUNSFELD: I wish Steve Squyres were here right now. I think he would argue with you. He has two of his children on Mars right now. I don't know. Jim, do you want to comment just a little bit about our program, what some of the next steps are that are already in place?

JIM GARVIN: Yeah. Well, I think maybe Steve would do it better but if you don't think we're exploring now, maybe we *don't* communicate that well. But I think—two rovers 270 days on another world wandering at 300 percent beyond expected lifetime is a new demonstration of that. Cassini alone is exploring at the highest order.

MEL AVERNER: I'd like to respond.

JIM GARVIN: But, let me finish. I mean, I can go on and on with the legacy of how we explore. It's just that, right now, a lot of people, perhaps in the public sector—and I can't speak for them because I'm a geek and work for NASA—but when I talk to them at hockey games and things where they don't always care what we do, they're stunned by what we're doing and how we're exploring. And how we've learned to go from people on the surface of the Moon as our agents of exploration, being our representatives, to machines being those agents. And we're doing that so many different ways. We're so diversified. In fact, if you ask corporate America and many of my colleagues there, they're stunned. "You're doing all that, with that portfolio? You're nuts!"

MEL AVERNER: I'd like to get back to the bar last night. Becky, was that your name, doing what I would have done—drinking at the bar? Suppose you were to go back to the bar and talk to those people and say, "Well, we are exploring. We have two robots on Mars doing terrific scientific things." Would they say, "Wow, that is great, but when are we going?"

JOHN GRUNSFELD: Absolutely, I agree. But just to give you the counterargument—and I don't know what the current number is, but there have been 13 billion hits on the NASA Web site of which three-quarters . . .

UNIDENTIFIED SPEAKER: That's a false number. It's not 13 billion people.

JOHN GRUNSFELD: No, no. I didn't say it was 13 billion people.

UNIDENTIFIED SPEAKER: I know, but that's the impression that it leaves.

JIM GARVIN: But there are well over 100 million unique IP addresses, maybe 250 million total. It's all around the world, predominantly the U.S., but all around the world. And, you know, you could argue about the numbers, but it is so much greater than any other Web site that it's phenomenal. There is interest there, and there's interest specifically because, I believe, that what we've done is we've put two human eyeballs on the surface of Mars. So people see what the rovers see and they think, "This is kind of what I would see when I get to go." Or, "When we send people, this is what they will see." And we want to do that.

QUESTION: I'm Nancy Ann Budden, Naval Postgraduate School and Lunar Planetary Institute in Houston. I want to build on some comments that were made by Joe Fuller and others about getting the word out and on some communications issues that Jim brought up. I joined Johnson Space Center's Exploration Office in '88 and I worked with a lot of you, Chris and Dale, on human exploration issues, and this was about the time that Bush '41 came out with his announcement that we were going back to the Moon and on to Mars.

One of the things that we neglected to do over the next 12 years, really, was put into place a communications plan. We all had great ideas. We had a lot of meetings. And now we have another opportunity with Bush '43 coming out with a much more reasonable, cost-rational plan and vision. And one thing I think we really need to do is put together a communications strategic plan, like a mission, and have a schedule and a budget and have somebody own that. Whether it's PAO [Public Affairs Office] through NASA Headquarters or whether it's an industry/NASA/university team. But we need to have a plan for that, that actually has someone own it, someone that's going to pay for it, and understand who are the advocates that we need to build. Obviously, there are communities we need to get to within NASA, of course. We need to get to the [Capitol] Hill. But we need to do it in an integrated, planned way with someone thinking about, okay—who are the first people we need to get to, and when and why, and how do we integrate this message? I nominate Keith Cowing to put together the message [laughter]. And, John, I think everyone would love for you to run the communications strategy idea since you're getting asked to do a lot of other things this morning and since you have a lot of spare time!

Anyway, I would like to see someone own that and put together a message that people agree with and actually stand behind, and make sure that it is consistent with our Commander in Chief's vision of the future for space exploration.

JOHN GRUNSFELD: That is absolutely a great comment. We've received that comment quite a lot, so we've actually heard that message and we've acted on it. Part of the transformation was to create a communications group, and we've linked the legislative and the public affairs and our external relations into one team so that we can help craft it. We were at the bar as well last night, talking about a budget, specifically, or an increased budget, line items, and management for public affairs as well. That's crucial, that we have to treat that as something that's very high-priority. But in the transformation, we've combined all of those for exactly the reason that you mentioned. Thank you.

QUESTION: David Gast. I'm the other student here for the school. The thing that I think everyone here is touching on, and building on some of the things that have just been said, is it is about communicating to the public. I think everyone in this room and most of the people watching NASA TV already know, kind of, the reasons that we want to go out there, what we hope to accomplish, where we hope to go, and understand the risks that are inherent to doing that. With this communication message, what we have to do is say to everyone else, the people that aren't in this room and aren't watching NASA TV, "This is where we want to go and this is why we want to go there. And, you know what? It's dangerous. Very likely, things are going to crash. Maybe people are going to die. But the people that are putting themselves on the line for that understand that and accept those risks for themselves and believe that the goal of what we're trying to accomplish is worth that risk." So, I think it's all these things.

We have to communicate the risk, yes, coupled with why we think the risks are worth taking. We can't just say, "We're going to do these great things, we're going to go to Mars, go to the Moon, and it will all be safe and happy and fun." Neither can we say, "It's dangerous to travel through space." We have to say all these things at the same time.

We talk about [that] the American people won't accept that something went wrong that we could have avoided. There's always one more thing we could have avoided had someone happened to think of it, had someone happened to see it. And I think they're willing to accept that if we're doing the best we can with what we have, there are always dangers there. And they're willing to accept that, again, if we communicate that to them in advance. Like I said, the people here all understand that. We need to take what we've talked about here and present that to American people.

JOHN GRUNSFELD: I think it's T. K. Mattingly who told us, "Success always has failure as its predecessor." He was more eloquent.

QUESTION: Keith Cowing. Thank you, Nancy, for the nomination. When you hear what I have to say, you may withdraw it. To the point of Web traffic—and you're right, I do Web sites for a living—citing Web numbers is so 1997, so Pathfinder. [laughter] Google does that traffic before lunch on Sunday. It's great to hear these numbers, but I could go write something in my room right now.

Drudge Report would pick it up and have a million hits by tonight. Big deal. The Web hit numbers are important. A lot of people are looking at NASA's Web sites. But we need to move on to other metrics. When a nine-year-old girl raises her hand at a Presidential visit and asks about space—things like that—then you know. When the late-night shows make different jokes about space—Jim Garvin has done yeoman's duty, going on *Letterman* and so forth. When you start to see this consciousness of space percolating up in other places... These numbers can be very deceiving. Anybody can generate hits. You've just got to look for other metrics. You've got to have a new metric every month. Just some advice from somebody who does this for a living.

UNIDENTIFIED SPEAKER: This sort of follows your point, Keith. But when the NASA crews come into the small town of Lander, Wyoming, to go on [National Outdoor Leadership School] courses, they often stay after and talk to the kids attending courses. And when those kids leave the room, they're changed. And following on your point, I believe, it's not about communicating to the public, it's about changing the public.

KEITH COWING: As the Administrator of NASA loves to say—it is this Jesuit thing he has—"one conversion at a time." It works. [laughter] It's self-propagating if you do it right.

JOHN GRUNSFELD: I should say that every time an astronaut leaves the school, they're changed as well.

QUESTION: Bill Clancy at NASA Ames. One concept that we haven't talked a lot about here that I found very useful as it relates to the public, and also inside, is the word sustainability. To me that's the most important word, I think, that's in our current vision. And I found it very useful to the shift from thinking about particular missions to the program. So, rather than just talking about mission risk, we have program risk. And we're talking about building competence and the ability to go places and so on.

I first understood this, I think, with Mars Polar Lander, where we didn't have the telemetry that we needed to give us the information for building the redesign that we needed. I think your example this morning is a beautiful example as well, of the investment that one can make to build tools that will give us a competence that we know we want to have [as] part of our tool kit. So, I think when we're articulating to ourselves what's our priority and our objective, it's the clear objective, maybe dates, and the sense of challenge. But it's all about sustainability, and we make decisions because we need to be here tomorrow. We're not going to climb Everest today, because just getting to Everest today is not our goal. We want to be able to climb again tomorrow.

JOHN GRUNSFELD: Anyone else? David.

DAVID HALPERN: Thank you, John. And one of the things we've learned—some of us knew before, but some others learned—that 96 percent of space needs to be explored and 96 percent of the ocean needs to be explored. One [space] has zero pressure and one [the ocean] has a very large pressure on the bottom. And then the question comes that the ocean definitely is a place to explore, for two reasons. One is [for] the creation of new knowledge, which is the same as what you're talking about for outer space. But inner space also has a well-recognized aspect of creation of wealth. I mean, a number of [beneficial] activities have always gone on in the ocean—and I don't mean just transportation, but subsurface as well—and new ones are coming along, like genomics, oceanography, things like that.

So, then comes the question. In the new, transformed NASA, the challenge, now, would be to make use of the fact that oceans—or inner space—require the same type of dedication and the same type of methodologies as are being used in exploration of outer space, and it's something that the new NASA might want to consider. And it's actually well-poised for that because all of the science now is in the Science Mission Directorate. Rather than in two different stovepipes, it's all in one. It's a comment, not a question.

JIM GARVIN: I'm really grateful for you for saying that because my new job at NASA, with the many hats, is, in fact, to try to integrate the inner and outer space exploration in this new vision. So I'm looking, as is Ghassem Asrar [NASA's Science Deputy Associate Administrator] and John, we're all looking for the connections. Because I think the point with a vision, with an objective, with some of these good points about program-thinking, which we've had in EOS for Earth science, we've had in the Mars program, we hope to have throughout our program—the Shuttle program—is an aspect of risk that I think is the one that right now strangleholds a lot of us. And that is risk of our own interpersonal management structures to get the job done.

And that, perhaps, is the genesis of the transformation, to get around some of those things. But, you know, when organizations grow old they become well-rooted in certain directions. And breaking roots, it's like taking a root off a redwood out there. I mean, it's going to stay three hundred feet tall, so you don't want to have it fall over. You want to have it move. And other than slime molds, most large plants don't move.

But I think that's the challenge. The ocean is an exploration frontier that will teach us about high-pressure environments and knowledge and all that, and some shared technologies could be trialed there in the name of science and exploration to good end. And, you know, it's rather ironic to me that a large fraction of the ocean exists at 100-bar pressure, which is the average surface pressure of the planet Venus. And, you know, lots of living stuff there. Interesting to think about.

JOHN GRUNSFELD: Thanks, Jim. We'll take one more.

QUESTION: George Tahue from NASA Headquarters. Listening to some of the comments here, an analogy is coming to my mind. If you're familiar with the paleontologist Stephen Jay Gould and his description of evolution as punctuated evolution, I think NASA is, as a government agency, going through an evolution, and we will continue to do so. Where we're going with this is going to take a very long time, but there are certain points where there will be punctuations that make great changes in very short amounts of time. And I think Apollo, that era, was one of those points. We may go through slower periods of time where we go through those changes. But here we're at another point where we may be at another one of those punctuations. And this new transformation that we're looking at isn't just rearranging the deck chairs. And it's something that we have to take internally and not just focus only on, why didn't the public understand what we're doing and how can we make them understand? It's something that we have to do over this long period of time, even internally.

When we had our transformation and the Office of Earth Science and the Office of Space Science came together, I was listening to some of the Earth science guys and saying, "Wow, you do that? That's cool!" Same reaction [as] at the bar. So, I'd like to charge all of us to try to take a lot of this internally and focus on those goals.

Another key thing we've heard here is to focus on the target. Stay on target. Protect and understand our planet. Search for life. Understand the limits of it, and recognize that humans and robots are the tools to do those goals. It's not just, "Get us there." It's not just, "Get the robots there." Focus on those [larger] goals. We'll have these punctuated evolutions where we have a grand target that we're looking for. And, in between, we'll have this balance that we keep going forward in trying to get that message to the public to understand that we, as an agency, have a role as a public function in our society. So, those are my thoughts.

QUESTION: I'm John Gaff from the Glenn Research Center. I think the Agency, while it does wonderful things—and I've been in it a long time, is not recognized by our society as critical to the survival of society. Nobody questions why you've got the State Department, nobody questions why you've got the Treasury, and nobody questions why you've got the Defense Department, or Agriculture, even. But for some reason, we have been unable, in my opinion, to transfer the knowledge that we are able to acquire for the future to being something critical for the survival of the economic success of the Nation. And for the long-term viability of the Nation.

Somehow, we need to start some mechanism—and maybe it's in the education programs, these outreach things—where we get more institutionalized as a recognized, long-term investment. Until that happens, we're always going to be at the margin, we're going to be at less than half a percent of the budget, and we're not going to be able to compete for the other critical needs of, "What's in it for me?" with the society. Thank you.

JOHN GRUNSFELD: I think that's a very good comment. I would like to point out that we're in relatively tough economic times right now, yet NASA is the only agency that's basically gotten an increase in its budget.

UNIDENTIFIED SPEAKER: Did it get one?

DISCUSSION

JOHN GRUNSFELD: Well, in the request, in the request. And even in the appropriations meetings, we've fared better than virtually all discretionary agencies. I think the issue is: We're still a discretionary agency.

UNIDENTIFIED SPEAKER: I'd like to kind of second that and say two things. You guys are in a really tough position, almost a harder position than you were in the Apollo era, because Apollo was something we all thought we needed to do. You guys are in the very, very hard position of deciding what we should do. We don't necessarily need to do anything, it doesn't look like. There's not an immediate and obvious need. But you guys can do lots of different things. I'd like to tack onto Dr. Halpern and Dr. Garvin's comments. I think you guys recognized this, but I'm not sure that the rest of the population does—one of the amazing things about the way NASA is exploring the new frontiers in space, and the way that ocean science explores our frontiers here on Earth, is that for the first time, I think, in human history, you've got the conservationist, the naturalist, the scientist, and the greedy capitalist wound up, in many cases, in one mind, in one human being. And you've got an organization that's already looking to protect resources that we can't even exploit yet or use yet.

I mean, does that seem strange to anybody else? That's new, folks! I mean, I think even more so than technologies, you guys can share lessons learned and organizational experiences based on how do we commercialize this thing, and how do we get benefit out of it as a people without destroying it for ourselves and our posterity? And perhaps that could be part of your public outreach program, because, for instance, look at the market for the Toyota Prius cars. It's huge! They're back-ordered, I don't know how long. Eight months back-ordered on the cars! You know? The public gives a darn about that kind of thing and you guys do it [balance benefits and conservation] every day. It's innate. It's part of your nature. That's important.

UNIDENTIFIED SPEAKER: Coupled with that and, again, talking about expanding the vision and explaining the risk, is that the vision we want to put out there is not just [that] we want to go back to the Moon and learn how to go to Mars. But I think it's a bigger vision than that. It's partially this and partially the thing that he's talking about. It's [that] we need to present both, this is the next step on which we are currently embarking, but also, this is a vision for the future that we hope to achieve by taking these steps. And that vision doesn't have to be perfect. It doesn't have to be exactly what we're going to arrive at. But it has to be a goal beyond just, you know, as great as the goal was to put a man on the Moon and bring him back to Earth. Why? Now why are we doing that?

And we've talked a lot about that, but I think that needs to be part of what would go out to the public, and what NASA thinks about internally, and each of us thinks about internally, in ourselves, as what is our long-term pictured goal that all these things are steps toward? And that goes for exploring the seas as well. You know, all these explorations are not just, I want to go to the bottom of the Marianas Trench. It's, I want to find out more about the Earth. I want to discover more about us as a species. I want to maybe discover things that will save our species or our Nation or whatever at some point in the future. So that needs to be a part under consideration as well.

JOHN GRUNSFELD: Okay, well, I think we're up to the end here. I just want to give all of you a big "thank you very much" for participating in this. I know I've learned a lot. I think we've all had a lot of good dialogue. I got a few too many action items, but they're very important ones and we will take that forward, back to NASA, and for those of us here from NASA, I hope you take that all out. I really want to encourage you again, though, as you leave here, to regard this as the start of a dialogue. There's no question that this is one we'll talk about sustaining. I think this dialogue will be sustained probably for all of human history as we push our frontiers, as we move out.

I'd like to bring Scott Hubbard, the Director of the Ames Research Center, to give us some closing comments.

CPSIA information can be obtained
at www.ICGtesting.com
Printed in the USA
LVHW020610130522
718627LV00009B/581